U.S. Department of Transportation
National Highway Traffic Safety Administration

DOT HS 811 497

July 2011

Independent Evaluation of the Driver Acceptance of the Cooperative Intersection Collision Avoidance System for Violations (CICAS-V) Pilot Test

DISCLAIMER

This publication is distributed by the U.S. Department of Transportation, National Highway Traffic Safety Administration, in the interest of information exchange. The opinions, findings, and conclusions expressed in this publication are those of the authors and not necessarily those of the Department of Transportation or the National Highway Traffic Safety Administration. The United States Government assumes no liability for its contents or use thereof. If trade names, manufacturers' names, or specific products are mentioned, it is because they are considered essential to the object of the publication and should not be construed as an endorsement. The United States Government does not endorse products or manufacturers.

REPORT DOCUMENTATION PAGE		Form Approved OMB No. 0704-0188
Public reporting burden for this collection of information is estimated to average 1 hour per response, including the time for reviewing instructions, searching existing data sources, gathering and maintaining the data needed, and completing and reviewing the collection of information. Send comments regarding this burden estimate or any other aspect of this collection of information, including suggestions for reducing this burden, to Washington Headquarters Services, Directorate for Information Operations and Reports, 1215 Jefferson Davis Highway, Suite 1204, Arlington, VA 22202-4302, and to the Office of Management and Budget, Paperwork Reduction Project (0704-0188), Washington, DC 20503.		
1. AGENCY USE ONLY (Leave blank)	2. REPORT DATE July 2011	3. REPORT TYPE AND DATES COVERED January 2008 – September 2010
4. TITLE AND SUBTITLE Independent Evaluation of the Driver Acceptance of the Cooperative Intersection Collision Avoidance System for Violations (CICAS-V) Pilot Test		5. FUNDING NUMBERS Inter-Agency Agreement HS-51A1 DTNH22-08-V-00017
6. AUTHOR(S) Mary Stearns and Lisandra-Garay Vega		
7. PERFORMING ORGANIZATION NAME(S) AND ADDRESS(ES) U.S. Department of Transportation Research and Innovative Technology Administration John A. Volpe National Transportation Systems Center Cambridge, MA 02142		8. PERFORMING ORGANIZATION REPORT NUMBER DOT-VNTSC-NHTSA-11-09
9. SPONSORING/MONITORING AGENCY NAME(S) AND ADDRESS(ES) John Harding U.S. Department of Transportation National Highway Traffic Safety Administration Washington, DC 20590		10. SPONSORING/MONITORING AGENCY REPORT NUMBER DOT HS 811 497
11. SUPPLEMENTARY NOTES		
12a. DISTRIBUTION/AVAILABILITY STATEMENT Document is available to the public from the National Technical Information Service www.ntis.gov		12b. DISTRIBUTION CODE
13. ABSTRACT (Maximum 200 words) This report documents the results of the independent evaluation's assessment of the driver acceptance of the Cooperative Intersection Collision Avoidance System limited to Stop Sign and Traffic Signal Violations (CICAS-V) system as tested during a pilot test in 2008. The purpose of this pilot test was to use naïve participants in on-road and test-track environments to assess the readiness and maturity of the CICAS-V for a large-scale field operational test. Data were evaluated from 87 naive drivers who were placed into equipped vehicles to navigate a two-hour prescribed route through ten stop-controlled and three signal-controlled equipped intersections. To ensure the data were sufficient to understand drivers' impressions of the warning, 18 drivers followed the on-road study with a test-track study. Overall, subjects were neutral to slightly satisfied with the CICAS-V. A similar trend was observed for their assessment of whether the system will increase their driving safety. The data suggest that subjects need to experience the system to be able to assess it fairly. This assessment of driver acceptance was limited by the scope of the pilot test as well as the expected low frequency of intersection violations.		
14. SUBJECT TERMS Cooperative Intersection Collision Avoidance System, CICAS-V, driver acceptance, pilot test, usability, alert timing, satisfaction rating, safety rating, endorsement, driver-vehicle interface		15. NUMBER OF PAGES 36
		16. PRICE CODE
17. SECURITY CLASSIFICATION OF REPORT Unclassified	18. SECURITY CLASSIFICATION OF THIS PAGE Unclassified	19. SECURITY CLASSIFICATION OF ABSTRACT Unclassified	20. LIMITATION OF ABSTRACT

NSN 7540-01-280-5500

Standard Form 298 (Rev. 2-89)
Prescribed by ANSI Std. 239-18
298-102

METRIC/ENGLISH CONVERSION FACTORS

ENGLISH TO METRIC | METRIC TO ENGLISH

LENGTH (APPROXIMATE)

English to Metric	Metric to English
1 inch (in) = 2.5 centimeters (cm)	1 millimeter (mm) = 0.04 inch (in)
1 foot (ft) = 30 centimeters (cm)	1 centimeter (cm) = 0.4 inch (in)
1 yard (yd) = 0.9 meter (m)	1 meter (m) = 3.3 feet (ft)
1 mile (mi) = 1.6 kilometers (km)	1 meter (m) = 1.1 yards (yd)
	1 kilometer (km) = 0.6 mile (mi)

AREA (APPROXIMATE)

English to Metric	Metric to English
1 square inch (sq in, in^2) = 6.5 square centimeters (cm^2)	1 square centimeter (cm^2) = 0.16 square inch (sq in, in^2)
1 square foot (sq ft, ft^2) = 0.09 square meter (m^2)	1 square meter (m^2) = 1.2 square yards (sq yd, yd^2)
1 square yard (sq yd, yd^2) = 0.8 square meter (m^2)	1 square kilometer (km^2) = 0.4 square mile (sq mi, mi^2)
1 square mile (sq mi, mi^2) = 2.6 square kilometers (km^2)	10,000 square meters (m^2) = 1 hectare (ha) = 2.5 acres
1 acre = 0.4 hectare (he) = 4,000 square meters (m^2)	

MASS - WEIGHT (APPROXIMATE)

English to Metric	Metric to English
1 ounce (oz) = 28 grams (gm)	1 gram (gm) = 0.036 ounce (oz)
1 pound (lb) = 0.45 kilogram (kg)	1 kilogram (kg) = 2.2 pounds (lb)
1 short ton = 2,000 pounds (lb) = 0.9 tonne (t)	1 tonne (t) = 1,000 kilograms (kg) = 1.1 short tons

VOLUME (APPROXIMATE)

English to Metric	Metric to English
1 teaspoon (tsp) = 5 milliliters (ml)	1 milliliter (ml) = 0.03 fluid ounce (fl oz)
1 tablespoon (tbsp) = 15 milliliters (ml)	1 liter (l) = 2.1 pints (pt)
1 fluid ounce (fl oz) = 30 milliliters (ml)	1 liter (l) = 1.06 quarts (qt)
1 cup (c) = 0.24 liter (l)	1 liter (l) = 0.26 gallon (gal)
1 pint (pt) = 0.47 liter (l)	
1 quart (qt) = 0.96 liter (l)	
1 gallon (gal) = 3.8 liters (l)	
1 cubic foot (cu ft, ft^3) = 0.03 cubic meter (m^3)	1 cubic meter (m^3) = 36 cubic feet (cu ft, ft^3)
1 cubic yard (cu yd, yd^3) = 0.76 cubic meter (m^3)	1 cubic meter (m^3) = 1.3 cubic yards (cu yd, yd^3)

TEMPERATURE (EXACT)

English to Metric	Metric to English
$[(x-32)(5/9)]$ °F = y °C	$[(9/5) y + 32]$ °C = x °F

QUICK INCH - CENTIMETER LENGTH CONVERSION

QUICK FAHRENHEIT - CELSIUS TEMPERATURE CONVERSION

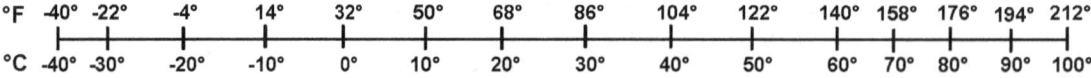

For more exact and or other conversion factors, see NIST Miscellaneous Publication 286, Units of Weights and Measures. Price $2.50 SD Catalog No. C13 10286

Updated 6/17/98

TABLE OF CONTENTS

EXECUTIVE SUMMARY vii
1. INTRODUCTION 1
   1.1. CICAS-V Description 1
   1.2. The CICAS-V Pilot Test 3
      1.2.1. Pilot Test Description 3
      1.2.2. Pilot Test Procedure 3
      1.2.3. Debriefing and Data Collection Instrument 4
2. DRIVERS PARTICIPATING IN THE CICAS-V PILOT TEST 6
   2.1. Subjects' Demographic Characteristics 6
3. INCIDENCE AND VALIDITY OF THE CICAS-V ALERTS 7
4. DRIVER ACCEPTANCE 11
   4.1. Overall Assessment of the CICAS-V 11
      4.1.1. Rating of Satisfaction with the CICAS-V 11
      4.1.2. Rating of Safety with the CICAS-V 12
      4.1.3. Suggestions for Improving the CICAS-V 13
   4.2. Usability of the CICAS-V 14
      4.2.1. Usability of the "Intersection Ahead" Display 14
      4.2.2. Usability of the "Running Red Light" and "Running Stop Sign"
      Alerts by Alert Modality 16
      4.2.3. Evaluation of Subjects' Perception of Alert Timing 18
      4.2.4. Self-Reported Alert Frequency Including Nuisance Warnings 19
   4.3. Endorsement of CICAS-V 21
5. SUMMARY 23
6. REFERENCES 25
Appendix A. Removal and Reclassification of Some Test Data 26

LIST OF FIGURES

Figure 1. Cabinet Containing Roadside Equipment at VTTI Smart Road Signalized Intersection 2

Figure 2. Steady Continuous Blue Advisory Intersection Ahead Display (left) and Flashing Red Visual Warning Display (right) 3

Figure 3. Driver Age and Gender by Post-Drive Questionnaire Format (79 Subjects).................. 6

Figure 4. Assignment of Subjects by their Experience with CICAS-V Alerts.............................. 8

Figure 5. Mean Post-Drive Questionnaire Responses to Satisfaction with "Running Stop Sign" Alert and to "Running Red Light" Alert by Driving Experience 12

Figure 6. Mean Post-Drive Questionnaire Responses about Perceived Increased Safety with "Running Stop Sign" Alert and to "Running Red Light" Alert by Driving Experience 13

Figure 7. Mean Post-Drive Questionnaire Responses Assessing the Location and Effectiveness of the "Intersection Ahead" Display by Driver Experience with CICAS-V Alerts.......................... 15

Figure 8. Mean Post-Drive questionnaire Responses Assessing the Distraction and Annoyance of the "Intersection Ahead" Display by Driver Experience with CICAS-V Alerts.......................... 15

Figure 9. Mean Post-Drive Questionnaire Responses for the Usability, Distraction, and Annoyance by Alert Modality for Subjects Who Had Received at Least One CICAS-V Alert .. 16

Figure 10. Mean Post-Drive Questionnaire Responses on Startling Quality and Effectiveness of CICAS-V Alerts for Subjects Who Had Received at Least One CICAS-V Alert....................... 17

Figure 11. Responses to Questionnaire Items about Brake Pulse Usability for Subjects Who Had Received at Least One CICAS-V Alert 18

Figure 12. Responses to Questionnaire Items about Alert Timing for Pseudo-Naturalistic and Smart Road Subjects 18

Figure 13. Responses to Questionnaire Items about Braking without Checking for Traffic for Pseudo-Naturalistic Subjects 19

Figure 14. Responses to Questionnaire Items about Their Experience with Alerts for Pseudo-Naturalistic Subjects 20

Figure 15. Responses to Questionnaire Items about Their Experience with "Running Stop Sign" Alerts for Pseudo-Naturalistic Subjects that Received at Least One Alert (any alert)................. 20

Figure 16. Post-Drive Questionnaire Responses to Inquiry about Willingness to Purchase the System by Subject Experience with CICAS-V Alerts... 21

Figure 17. Post-Drive Questionnaire Responses to Inquiry as to Amount of Money Willing to Pay to Purchase by Subject Experience with CICAS-V Alerts 22

LIST OF TABLES

Table 1. Drivers' Exposure to CICAS-V Alerts 9

Table 2. Comparison of the Subjects' Perception of an Intersection Violation with Actual Violation 10

Table 3. Suggestions for Improving CICAS-V 14

Table 4. Data Removed from Independent Evaluation Data Set (8 Subjects) 26

Table 5. Data Reclassified in Independent Evaluation Data Set (4 Subjects) 26

LIST OF ACRONYMS

CAMP	Crash Avoidance Metrics Partnership
CICAS-V	Cooperative Intersection Collision Avoidance System for Violations
DVI	Driver-Vehicle Interface
FOT	Field Operational Test
OBE	On Board Equipment
RSE	Roadside Equipment
VTTI	Virginia Tech Transportation Institute

EXECUTIVE SUMMARY

The Volpe National Transportation Systems Center of the U.S. Department of Transportation's (U.S. DOT) Research and Innovative Technology Administration performed an independent evaluation of the driver acceptance of the Cooperative Intersection Collision Avoidance System for Violations (CICAS-V). This effort was part of the U.S. DOT Intelligent Transportation System CICAS-V initiative to accelerate the introduction of emerging technology to address crashes at intersections. This evaluation was based on subjective data collected from a pilot test of the full implementation of the CICAS-V in anticipation of eventually conducting a field operational test. The National Highway Traffic Safety Administration managed this project in cooperation with the Crash Avoidance Metrics Partnership that developed and built the CICAS-V and the Virginia Tech Transportation Institute that conducted the pilot test.

Data were evaluated from 87 naive drivers who were placed into CICAS-V equipped vehicles to navigate a two-hour prescribed route through equipped intersections in the New River Valley region of Virginia. During the prescribed route, drivers crossed ten stop-controlled and three signal-controlled intersections making a variety of turn maneuvers through each for a total of 52 intersection crossings. To ensure the data were sufficient to understand drivers' impressions of the warning, 18 drivers followed the on-road study with a test-track study. The test-track study used a ruse that required drivers to perform a variety of in-vehicle tasks while driving. A distraction task was delivered at a carefully controlled point near a signalized intersection so that drivers were not looking at the traffic signal during the phase change. A CICAS-V warning was presented and the drivers' responses to the warning were recorded. Three modalities were used to warn subjects: visual, auditory (speech) and haptic (brake pulse).

Overall, subjects were neutral to slightly satisfied with the CICAS-V. A similar trend was observed for their assessment of whether the system will increase their driving safety. Subjects suggested enhancing the conspicuity of the visual blue "intersection ahead" display and using a more eye catching color or intermittent flashing to improve awareness of it. Subjects who received alerts rated the speech alert and brake pulse very easy to detect and effective in warning them about potential intrusion into an intersection. By contrast, subjects were neutral to slightly negative about the ease of detecting and the warning effectiveness of the "intersection ahead" display and the "running red light/stop sign" display. Subjects rated the timing of the "Running Red Light" and "Running Stop Sign" CICAS-V alerts as about right and maybe a little late. In general, between one in seven and one in five subjects said they could not identify the source of the alert and that they did not receive one when they felt it was needed. Almost two in five subjects said that they received a "Running Stop Sign" alert that was appropriate. Subjects who had received an alert slightly agreed that they would purchase the CICAS-V, cost aside. Subjects who did not receive an alert said that they definitely would not purchase the system. The data suggest that subjects need to experience the system to be able to assess it fairly. There is the possibility that the subjective definition of an intersection violation may not map exactly to the operational definition embodied in the CICAS-V. This was shown by the mismatch between the CICAS-V definition of violation versus subjective assessments. Finally, this assessment of the driver acceptance of the CICAS-V system was limited by the scope of the pilot test (i.e., subjects driving a prescribed two-hour route pilot test and a subset selected for the test track experience) as well as the expected low frequency of intersection violations.

1. INTRODUCTION

This report presents the results of the Independent Evaluation's assessment of the driver acceptance of the Cooperative Intersection Collision Avoidance System limited to Stop Sign and Traffic Signal Violations (CICAS-V) system as tested during the pilot test in 2008. The U.S. Department of Transportation established the CICAS-V program as one of the Intelligent Transportation System initiatives. The CICAS-V initiative is intended to work cooperatively with industry partners to accelerate the introduction of emerging technology to address crashes at intersections. As part of developing the CICAS-V initiative, the National Highway Traffic Safety Administration managed a pilot test of the full implementation of the CICAS-V in anticipation of eventually conducting a field operational test (FOT) of this technology to improve safety at intersections. The purpose of this pilot test is to use naïve participants in on-road and test-track environments to assess the readiness and maturity of the CICAS-V and to ensure the FOT will achieve its goals. The CICAS-V was developed by the Crash Avoidance Metrics Partnership (CAMP), a consortium of five automakers that includes Ford, General Motors, Mercedes-Benz, Toyota, and Honda. The pilot test was conducted by the Virginia Tech Transportation Institute (VTTI). The Volpe National Transportation Systems Center (Volpe Center) of the U.S. Department of Transportation's Research and Innovative Technology Administration performed the independent evaluation.

1.1. CICAS-V Description

The CICAS-V is intended to assist drivers to avoid intersection crashes. As described by the system developers..."The driver is issued a warning if the equipment in the vehicle determines that, given current operating conditions, the driver is predicted to violate the signal in a manner which is likely to result in the vehicle entering the intersections" [1].

The CICAS-V equipment is installed in On Board Equipment (OBE) within the test vehicle as well as on designated intersections in roadside equipment (RSE). The RSE is typically housed in a cabinet located near the equipped traffic control device (Figure 1). A central component of the RSE is the Wireless Access in Vehicular Environment radio, which transmits data regarding the intersection geometry and signal timing to CICAS-V equipped vehicles. This information, in combination with positioning data, allows an algorithm within the vehicle to determine if a violation is likely to occur. If the algorithm predicts a violation will occur, the OBE issues a warning to the driver through a Driver-Vehicle Interface (DVI). The OBE controls the three DVI modalities in the CICAS-V equipped vehicle, i.e., auditory (speech), visual, and haptic (brake pulse). The DVI has three states: 1) an inactive state when the vehicle is not approaching an equipped intersection; 2) a full flashing red violation warning mode that encompasses a "single stage" activation of the visual, auditory, and haptic alerts; and 3) a visual-only indication (the steady continuous blue advisory display) when the vehicle is approaching an equipped intersection [2].

Figure 1. Cabinet Containing Roadside Equipment at VTTI Smart Road Signalized Intersection

The auditory alert is sent through the on-board line-out jack to an amplified speaker. The speaker is located in the dashboard above the instrument cluster and independently of the vehicle's sound systems. The auditory alert consists of a female voice stating "Stop Light" or "Stop Sign", presented at 72.6 dBA out of the front speakers, measured at the location of the driver's head.

The visual alert was displayed by icons on the dash-mounted display positioned at the vehicle centerline near the windshield (Figure 2). The visual display was 11.6 mm (0.46 inches) high and 11.6 mm (0.46 inches) wide. The visual component was designed intentionally to be a secondary or supplemental part of the display while the brake pulse and speech tone are the primary means of conveying urgency and the need for action.

The haptic brake pulse was triggered immediately before the onset of the visual and auditory warnings, such that the deceleration would reach about 0.1g at approximately the same time as the visual and auditory warning onset. The total pulse duration was about 600 milliseconds. The deceleration produced by the brake pulse peaked at around 0.25g and was reached between 0.25 and 0.35 second after the onset of the visual and auditory warnings.

Figure 2. Steady Continuous Blue Advisory Intersection Ahead Display (left) and Flashing Red Visual Warning Display (right)

1.2. The CICAS-V Pilot Test

The pilot test provides a way to obtain initial feedback about the likely level of driver acceptance of the CICAS-V. The objective is to describe driver acceptance of the CICAS-V alerts overall and by type (Running Stop Sign or Running Red Light alerts) to provide insight into drivers' initial experience with the CICAS-V as well as identify any driver acceptance issues.

To support the driver acceptance goal of the independent evaluation, the Volpe Center collaborated with CAMP and VTTI to create the pilot test post-drive questionnaires. As the conductor of the pilot test, VTTI recruited the test subjects and instructed them to drive a prescribed route. The route included multiple passes of three CICAS-V equipped signalized and ten stop sign-controlled intersections. VTTI subsequently selected a subset of the pilot test subjects to participate in a follow-on study on the Smart Road to monitor their response to a surprise signal change at the CICAS-V equipped test track intersection.

1.2.1. Pilot Test Description

VTTI recruited the pilot test subjects from the Blacksburg and Christiansburg, Virginia area using newspaper ads, flyers, word of mouth as well as the VTTI subject data base. Interested people contacted VTTI and underwent telephone screening to eliminate candidates outside the designated age categories (18-30, 35-50, 55+), with certain health conditions, who did not possess a current license, used certain medications or had prior driving under the influence convictions. VTTI recruited 93 subjects to participate. During the course of the pilot test, VTTI reported that system failures resulted in a loss of data and there were usable data from 87 drivers.

1.2.2. Pilot Test Procedure

Upon arriving at VTTI, each subject was first asked to fill out consent forms and background information questionnaires. VTTI staff then gave each subject a static orientation to the vehicle pointing out and demonstrating its features covering the ignition procedure, HVAC system and seat adjustments. They also gave each subject a number to call in an emergency. The CICAS-V pilot test vehicles were two identically equipped 2006 Cadillac STS that had a forward collision warning system; a head-up display presenting vehicle speed on the windshield, and a backing aide. VTTI reported that they did not emphasize the CICAS-V more than the other safety features to make the subjects think that they were in a study to evaluate a number of safety systems.

VTTI instructed each subject to drive the CICAS-V equipped vehicle unaccompanied through a prescribed two-hour, 36-mile route, crossing 17 (3 signalized and 14 stop-controlled) CICAS-V equipped intersections. Ten of the 14 stop-controlled intersections and the three signal-controlled intersections were selected for the pilot test yielding a total of 13 intersections examined in the pilot test [2]. Each subject traversed each intersection many times from different directions to accumulate multiple crossing data from through, right-turn and left-turn maneuvers. The route included a total of 20 maneuvers at signal-controlled intersections and 32 maneuvers at stop-controlled intersections; the CICAS-equipped intersections were a subset of this total number of intersections on the route. Subjects were directed along the route by information from an auditory navigation device. VTTI reported that they evaluated CICAS-V alerts that occurred when the vehicle was traveling at a speed of 10 mph (16 km/h) or more because the algorithms were not designed to alert at speeds less than 10 mph [2].

A subset of the subjects was asked to participate in a follow on Smart Road test track study. These subjects drove on the Smart Road test track accompanied by an experimenter and were distracted on their final intersection approach to trigger the CICAS-V alert. The distraction protocol used in the Smart Road study was administrated as discussed in Perez et al. [3].

1.2.3. Debriefing and Data Collection Instrument

After completing the two-hour route, a VTTI staff member met each subject to lead them indoors to complete the post-drive questionnaire. At the same time, a second VTTI staff member downloaded data from the returned vehicle to obtain information about the driver's alerts.

VTTI developed three versions of the post-drive questionnaire to match the subjects' experience with CICAS-V alerts. A subset of the questions was repeated across versions where appropriate. If a subject did not receive any alerts when they drove the two-hour route and were not selected subsequently to participate in the additional Smart Road test, they were asked to complete the "No Alert" post-drive questionnaire. The "No Alert" post drive questionnaire includes questions about the "intersection ahead" display, endorsement of CICAS-V, perceived violation experience, and an open-ended question asking for suggestions to improve the CICAS-V.

A second version of the questionnaire that VTTI referred to as "Pseudo-Naturalistic" was given to subjects who had experienced at least one alert during their drive on public roads. A subgroup of these drivers was selected to participate in the additional Smart Road test track study. This subgroup did not complete their post-drive "Pseudo-Naturalistic" questionnaire until after they completed the test track study. The "Pseudo-Naturalistic" questionnaire includes the same questions that were in the "No Alert" questionnaire. In addition, the "Pseudo-Naturalistic" questionnaire contains questions asking about subjects' overall impressions of the "Running Red

Light" and "Running Stop Light" alerts; their perception of each of the three warning modalities (speech, haptic, visual display), and their perception of the reliability of the alerts.

VTTI labeled the third version of the post-drive questionnaire the "Smart Road" version. Subjects who only experienced an alert during the Smart Road test track study completed the "Smart Road" questionnaire. The "Smart Road" questionnaire includes questions from the "Pseudo-Naturalistic" questionnaire but omitted questions about overall impression with the "Running Stop Sign" alert and perception of the reliability (misses, correct and false alerts) of the "Running Red Light" and "Running Stop Light" alerts.

2. DRIVERS PARTICIPATING IN THE CICAS-V PILOT TEST

VTTI reported on 87 drivers who participated in the CICAS-V Pilot Study. The independent evaluator examined the data from each of these subjects and reformatted the data set to remove eight subjects who provided incomplete data. Appendix A provides the detailed reason for the elimination of these eight subjects from the data set. As a result, the independent evaluation reports on data from 79 subjects. In addition, the independent evaluator reclassified four subjects and the reason and reclassification action are enumerated in Appendix A by subject identification number.

2.1. Subjects' Demographic Characteristics

VTTI made an effort to recruit equal number of males and females on each of three age groups to participate in the CICAS-V pilot test. However, the data collection procedure required that each subject be administered one of the three post-drive questionnaires that matched their experience with alerts during their drive. The procedure of assigning subjects to one of the three versions of the post-drive questionnaire disrupted the initial proportionate distribution of subjects by gender and age. The relatively small number of subjects made it impractical to partition the data any further than by drivers' experience with alerts. As a result, the independent evaluation does not use gender and age as a basis for sorting subject responses to questionnaire items.

Figure 3 shows the demographic characteristics of the subjects sorted by the post-drive questionnaire format they answered. More than half (54%) of the subjects in the pilot test did not experience a CICAS-V alert, including 51 percent of the women and 54 percent of the men. Younger and older subjects were less likely to have experienced an alert than the middle-age subjects; 64 percent of the age 18-30 subjects did not experience any alert and 52 percent of the subjects age 55 and over did not experience any alert.

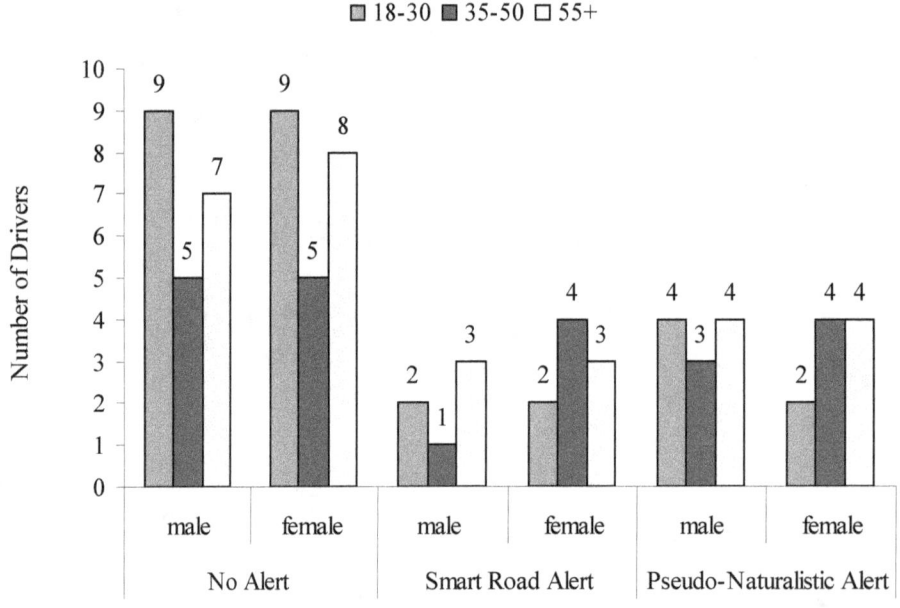

Figure 3. Driver Age and Gender by Post-Drive Questionnaire Format (79 Subjects)

3. INCIDENCE AND VALIDITY OF THE CICAS-V ALERTS

Figure 4 shows how the 79 subjects are distributed by their experience with CICAS-V violation alerts. Thirty six drivers (46%) received at least one violation alert at a CICAS-V stop sign-controlled or signalized intersection. This includes the 15 drivers who received an alert on the Smart Road test track and the 21 drivers (20 drivers on the public road and one driver on both the public and the Smart Road) who received at least one alert during their pseudo-naturalistic drive.

Thirty three (of the 36 subjects) or 92 percent of the subjects receiving an alert received at least one valid alert and five subjects (14%) received at least one invalid alert. Almost all the valid alerts were for stop signs. There was one valid alert for a traffic signal. In total, combining the on-road alerts (pseudo-naturalistic) and the Smart Road signal alerts, the 36 subjects who received at least one valid alert experienced 17 valid stop sign alerts and 16 valid signal alerts. Three subjects experienced only invalid alerts at signalized intersections.

The independent evaluation used VTTI's assignment of alerts as valid or invalid. VTTI's data reduction staff made the validity assignment ex post facto by reviewing video of episodes when alerts were issued as well as reviewing the DVI status, the vehicle's approach phase, the brake status, the distance to the stop bar, the correctness of the intersection ID, the longitudinal acceleration to confirm brake pulse activation, vehicle speed as well as several other factors [2].

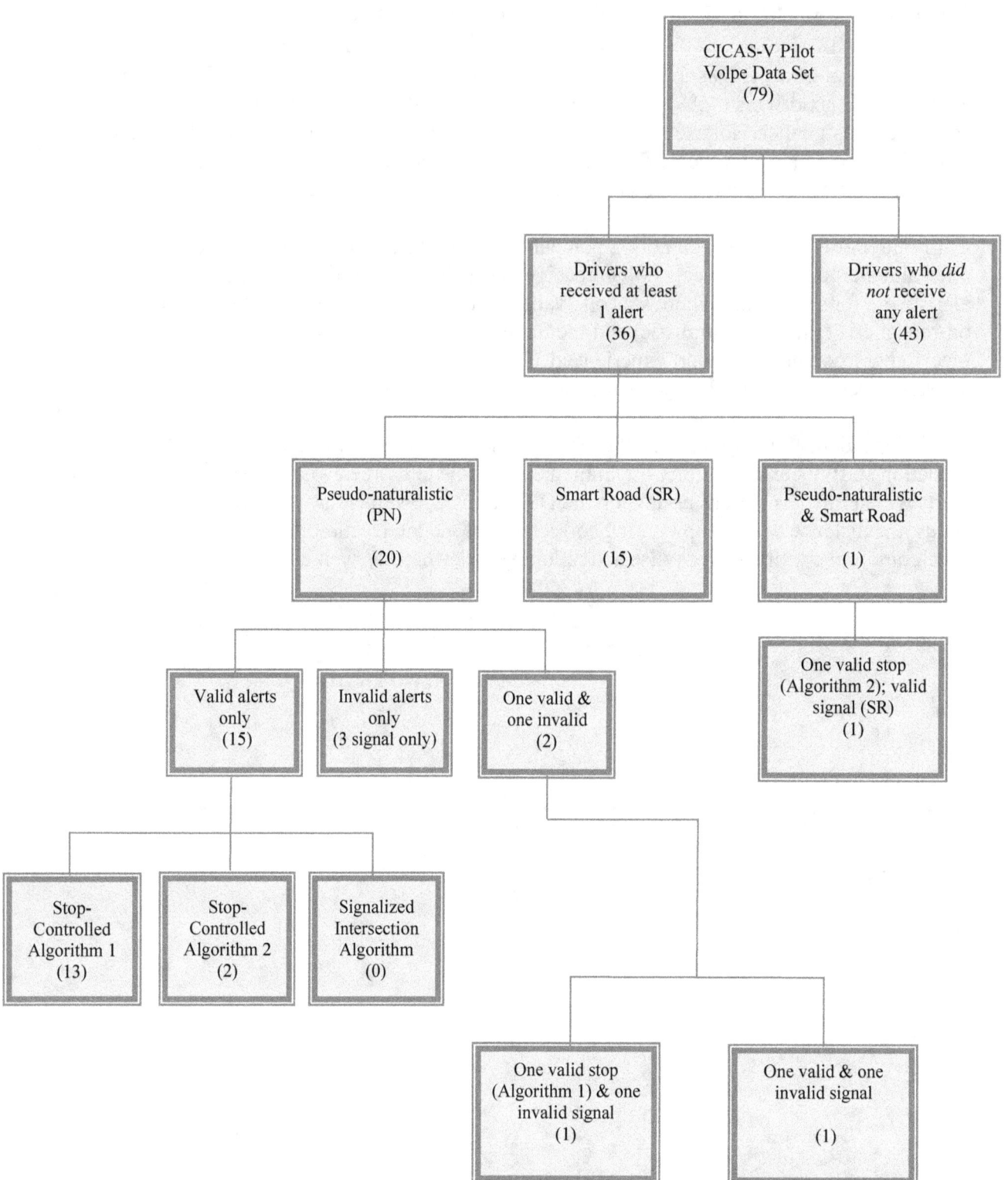

Figure 4. Assignment of Subjects by their Experience with CICAS-V Alerts

VTTI revised the algorithm used to trigger CICAS-V alerts at stop sign intersections during the pilot test. During the first several days of the field test of the CICAS-V, Stop-Controlled Algorithm 1 triggered many alerts at stop sign intersections. Fifteen of the 79 subjects or 19 percent approached a stop-controlled intersection driving with Stop-Controlled Algorithm 1. Fourteen of these drivers experienced a total of 50 valid alerts. VTTI replaced Stop-Controlled Algorithm 1 with Stop-Controlled Algorithm 2 for the remaining 64 subjects (81%)[1].

When VTTI downloaded data from test vehicles immediately following each subject's drive, they noticed that the fifteen drivers driving the test vehicle with Stop-Controlled Algorithm1 had alerts clustered at 5 CICAS-V equipped stop signs which were all on an uphill grade. Because Stop-Controlled Algorithm1 considered brake status in determining when to issue an alert, it issued alerts because drivers used the brake later than predicted due to the uphill grade. As a result, VTTI installed an alternative Stop-Controlled Algorithm 2 that monitored the deceleration level of the vehicle instead of the brake pedal status to determine whether to issue an alert. Consequently, the incidence of alerts at stop signs declined sharply using the Stop-Controlled Algorithm 2. Three of these 64 drivers or five percent received three violation alerts.

There were no changes to the traffic signal algorithm during the pilot test. The 79 subjects received a total of six alerts at signalized intersections. One alert was valid and the remaining five traffic signal alerts were invalid. Table 1 shows that almost one fifth (19%) of the pilot test subjects experienced Stop-Controlled Algorithm 1. They are the subset of subjects who experienced most of the valid alerts. After Stop-Controlled Algorithm 2 was introduced, the number of alerts decreased markedly and only five percent of the subjects received a valid alert with Stop-Controlled Algorithm 2. A much smaller proportion of the subjects, about one percent, received a valid traffic signal alert and eight percent received invalid traffic signal alerts.

Table 1. Drivers' Exposure to CICAS-V Alerts

	Number of Drivers by CICAS-V Algorithm	Number of Drivers Who Received a Valid CICAS-V Alert [2]	Number of Drivers Who Received an Invalid CICAS-V Alert	Mean Number of Crossings for Drivers Who Did Not Receive a CICAS-V Alert	Mean Number of Crossings for Drivers Who Received a CICAS-V Alert	Total Crossings (all drivers)	CICAS-V Alert Rate (Total Valid alerts/ Total number of crossings)
Stop-Controlled Algorithm 1	15 (19%)	14 (93%)	0	18	34	493	0.028
Stop-Controlled Algorithm 2	64 (81%)	3 (5%)	0	30	34	1448	0.002
Traffic Signal Algorithm	79 (100%)	1 (1%)	5 (8%)	16	19	1052	0.001

---

[1] The Stop-Controlled Algorithm 1 considered brake status when determining whether the driver should receive violation alert. Stop-Controlled Algorithm 2 did not rely on brake status to suppress the CICAS-V warning.
[2] One driver received once valid stop (Algorithm 1) and one invalid signal. One driver received one valid and one invalid signal.

It is worth noting that, despite use of an intersection violation warning system, subjects were more likely to subjectively perceive themselves as violating intersections that actually transpired. Subjects were asked for their perception of how often they thought they came close to violating a stop sign or traffic signal.

Table 2 compares the subjects' perception of their intersection violations against the violation alerts issued by the CICAS-V. Almost two thirds of the subjects thought that they violated an intersection compared to the less than half, 46 percent, that actually did incur an intersection violation alert from the CICAS-V.

Table 2. Comparison of the Subjects' Perception of an Intersection Violation with Actual Violation

	Number of Drivers that Violated or Came Close to Violate Signal or Stop Sign (Self Reported)			Number of Drivers that *Did not* Violate nor Came Close to Violate Signal or Stop Sign (Self Reported)	Total
	Stop	Signal	Both		
Number of Drivers that Received *at least one* Violation Alert	13	12	2	9	36 (46%)
Number of Drivers that *Did not* Received a Violation Alert	11	11	2	19	43 (54%)
Total	24	23	4	28	79
	51 (65%)			28 (35%)	79 (100%)

4. DRIVER ACCEPTANCE

The CICAS-V technology is intended to help drivers to avoid or minimize the severity of crashes occurring at intersections with traffic control devices by warning them when they are likely to violate the intersection's stop sign or traffic signal. It is useful to examine whether and how well the subjects in the pilot test accepted the CICAS-V they experienced during their two-hour drive on the prescribed route. The independent evaluation focuses on the components of acceptance for drivers which include an overall assessment of the CICAS-V, evaluation of aspects of usability, and endorsement of use.

The three versions of the post-drive questionnaire have some items in common. The three versions asked for drivers' opinions about the "intersection ahead" display, experience driving the test vehicle at red lights and stop signs, interest in purchasing the CICAS-V system, and for suggestions to improve the system. Two of the three questionnaires contained different subsets of questions to capture drivers' experience with CICAS-V alerts.

4.1. Overall Assessment of the CICAS-V

Drivers' subjective assessment of CICAS-V was examined in several ways including their responses to questionnaire items about satisfaction and safety as well as their suggestions for improvement.

4.1.1. Rating of Satisfaction with the CICAS-V

The following ratings of satisfaction with the CICAS-V were obtained from drivers who had experienced at least one alert. The only drivers who were asked about their satisfaction with the "Running Stop Sign" alert were in the pseudo-naturalistic group. Almost all of them (17 out of 21) had experienced at least one of these alerts. Their mean score for satisfaction shows that these subjects were slightly satisfied with the "Running Stop Sign" alert (top bar on Figure 5).

The Smart Road and Pseudo-Naturalistic subjects who had received a "Running Red Light" alert were asked to rate how satisfied they were with the CICAS-V. Figure 5, bottom bars, shows that the subjects who had the Smart Road experience, all of whom experienced a "Running Red Light" alert, were definitely satisfied with the "Running Red Light" alert. The subjects in the pseudo naturalistic category, although they had received alerts most of which were for running stop signs, had much less frequent experience with the "Running Red Light" alert and this is confirmed by their slightly better than neutral responses to this question.

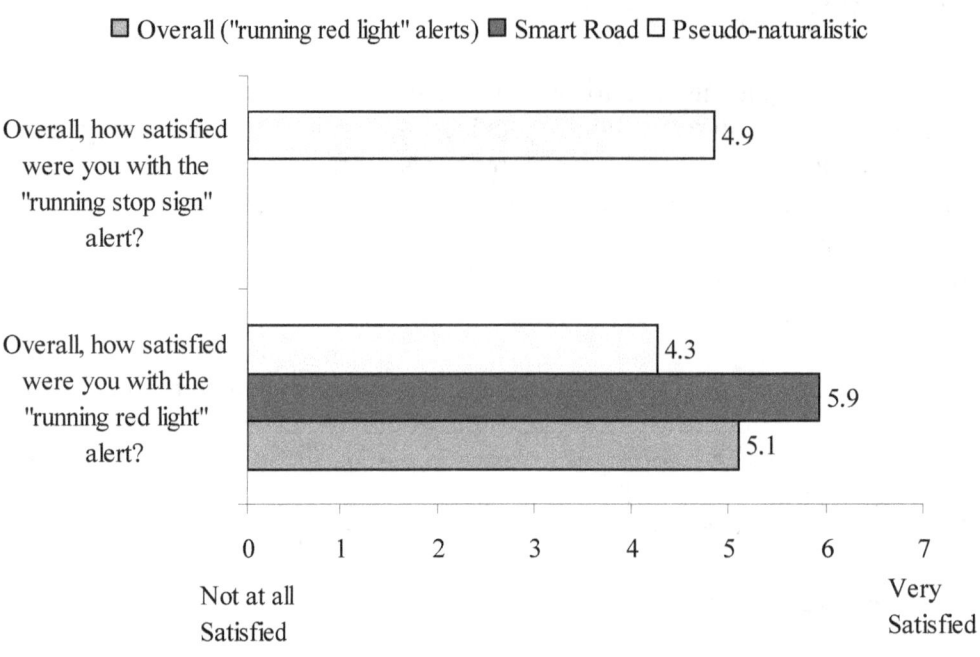

Figure 5. Mean Post-Drive Questionnaire Responses to Satisfaction with "Running Stop Sign" Alert and to "Running Red Light" Alert by Driving Experience

4.1.2. Rating of Safety with the CICAS-V

Ratings of perceived safety with the CICAS-V were obtained from subjects who had experienced at least one alert (36 subjects). Subjects who received a "Running Stop Sign" alert were slightly positive as to whether this alert will increase their safety (Figure 6). The only subjects who were asked about their safety with the "Running Stop Sign" alert were in the pseudo-naturalistic group.

Subjects who received an alert on the Smart Road strongly agreed that the "Running Red Light" alert will increase their safety. Subjects who received a "Running Red Light' alert during the pseudo-naturalistic study slightly agreed that this alert will increase their safety. Only five drivers experienced "Running Red Light" alerts during the pseudo-naturalistic study and most of these alerts were invalid.

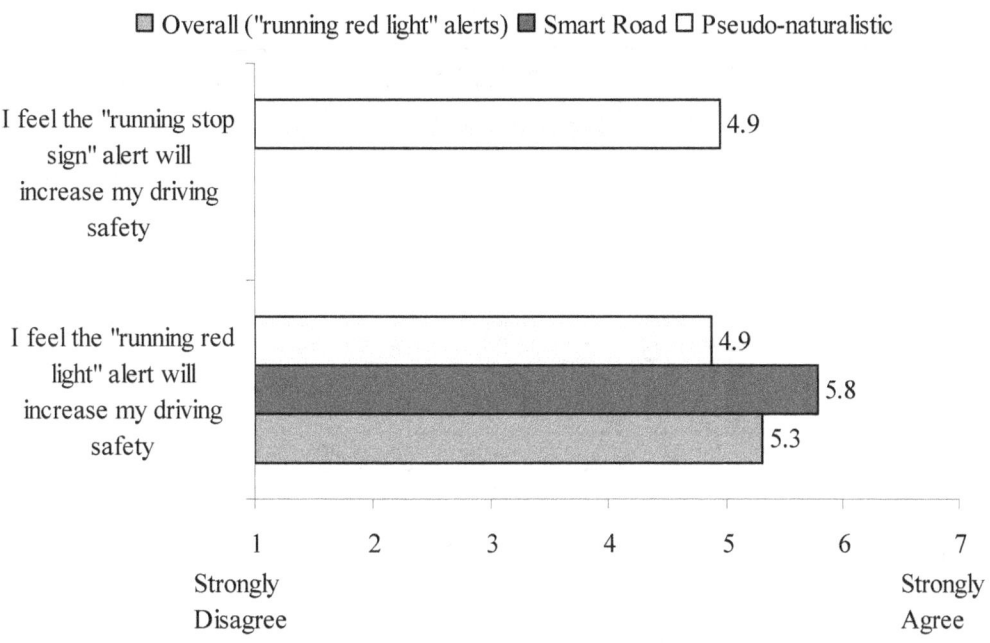

Figure 6. Mean Post-Drive Questionnaire Responses about Perceived Increased Safety with "Running Stop Sign" Alert and to "Running Red Light" Alert by Driving Experience

4.1.3. Suggestions for Improving the CICAS-V

Subjects were asked for their suggestions to improve the CICAS-V and more than four fifths (83%) provided them. The most frequently mentioned suggestions were to make the intersection warning system more conspicuous and to change and/or improve the color of its visual display (Table 3). Subjects who had not received any alerts were more likely to mention that they did not notice the system or even had forgotten what it did. It should be noted that the vehicle they drove had a number of new features such as a windshield display of the vehicle speed, a forward collision warning system, and a navigation system which may have divided the subjects' attention. In addition, VTTI reported that it did not single out the CICAS-V for attention during their orientation procedure and discussed CICAS-V as one of several safety features on the test car.

Table 3. Suggestions for Improving CICAS-V

"Do you have any suggestions for improving the intersection warning system?"	Alerts	
	0	1 or more
Topics	Frequency	
Did not notice (inconspicuous, forgot what it meant, not obvious)	18	7
Color (brighter, eye catching color, flashing light, poor contrast, sunlight washout, sunglasses obscured, dim)	13	7
Location (closer to driver, on windshield with mph or near speedometer)	6	5
Sound (add alerts, tone that can be turned on or off vs. for alert-- voice added little, too long; couldn't understand, unclear,)	5	7
Brake pulse (too aggressive, startling, effective, alarming, annoying		4
Timing (needed sooner, not issued in time, too soon when accelerating uphill)	2	4
Works well (interested in purchasing if it prevents running a light)	2	3
Size (make larger)	1	4
Offered no suggestions (good system)	1	4
Inconsistent operation (failed to illuminate for stop sign)	1	1

There were instances of inconsistent operation. VTTI reported that six drivers each received one invalid signalized intersection violation warning. VTTI reports that the CICAS-V was enabled 96 percent of the time for stop-controlled and signalized intersections. However, the stop-controlled intersections were more likely to be disabled and for longer durations. Three quarters of the drivers (75%) experienced a total of 239 disabled periods at the signalized intersections compared to 93 percent of the drivers experiencing a total of 628 disabled periods at stop-controlled intersections. Subjects experienced disabled stop-controlled intersections three times as often as they did a disabled signalized intersection, or 10 versus 3.7 times. This supports that last entry in Table 3 where a subject made a suggestion to make the display illuminate more consistently for equipped stop signs.

4.2. Usability of the CICAS-V

In the subsequent sections, driver acceptance scores are presented by driver experience with CICAS-V violation warnings (i.e., drivers that received at least one violation warning versus those who did not).

4.2.1. Usability of the "Intersection Ahead" Display

All three post-drive questionnaires asked subjects to rate the usability of the "intersection ahead" display in term of its size, location, detectability, and effectiveness in warning them about an upcoming CICAS-V intersection. Figure 7 shows the responses categorized by the subjects' experience with CICAS-V alerts. Subjects who had received an alert had a neutral opinion about the appropriateness of the size and location of "intersection ahead" display. These drivers were also neutral as to whether the display was easy to detect and effective in letting them know that the intersection warning system had detected an intersection ahead. Subjects who had not experienced an alert tended to rate the size of the "intersection ahead" display as inappropriate, slightly disliked its location, thought it was difficult to detect the display, and were slightly negative as to whether this display was effective in warning them about the intersection ahead.

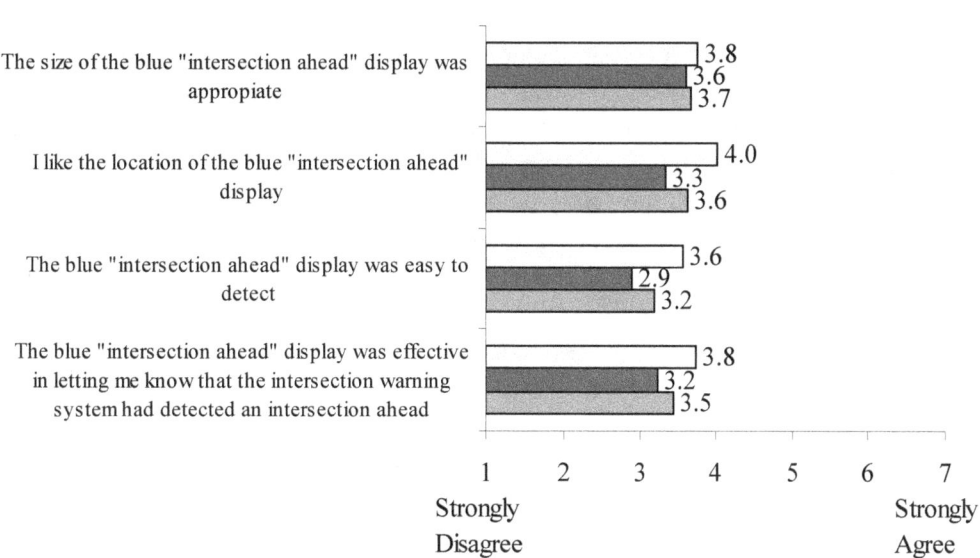

Figure 7. Mean Post-Drive Questionnaire Responses Assessing the Location and Effectiveness of the "Intersection Ahead" Display by Driver Experience with CICAS-V Alerts

Subjects were asked for their opinion as to whether they thought the "intersection ahead" display was distracting or annoying. Figure 8 shows the rank order of their mean responses to these items by driver experience with CICAS-V alerts. The subjects had similar ratings and did not rate the "intersection ahead" display as distracting or annoying.

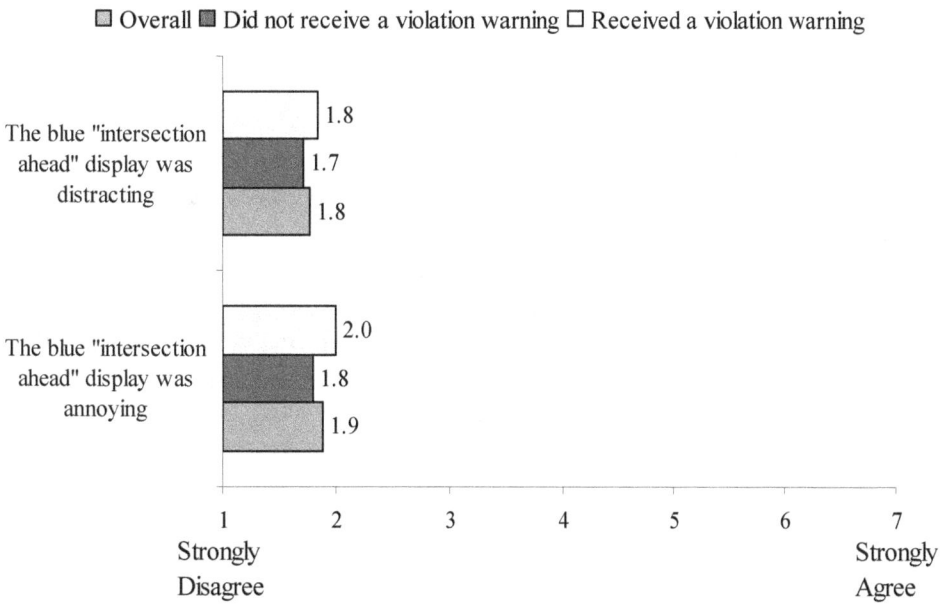

Figure 8. Mean Post-Drive questionnaire Responses Assessing the Distraction and Annoyance of the "Intersection Ahead" Display by Driver Experience with CICAS-V Alerts

4.2.2. Usability of the "Running Red Light" and "Running Stop Sign" Alerts by Alert Modality

The subjects who had received at least one alert (36 in Figure 4) were asked for their opinions about the usability, distraction, and annoyance of the alert components, (i.e., intersection ahead and running red light/stop sign display, speech alert and brake pulse). Figure 9 shows the results for by each alert modality by usability features including distraction, annoyance, detection and effectiveness of conveying information. These subjects agreed that the speech alert and the brake pulse were easy to detect. They also agreed that the speech alert and brake pulse were effective in letting them know that they may be about to run a red light or stop sign. Drivers were neutral as to whether the running red light/stop sign visual display was effective. None of the sensory modalities of the CICAS-V alerts were considered distracting or annoying.

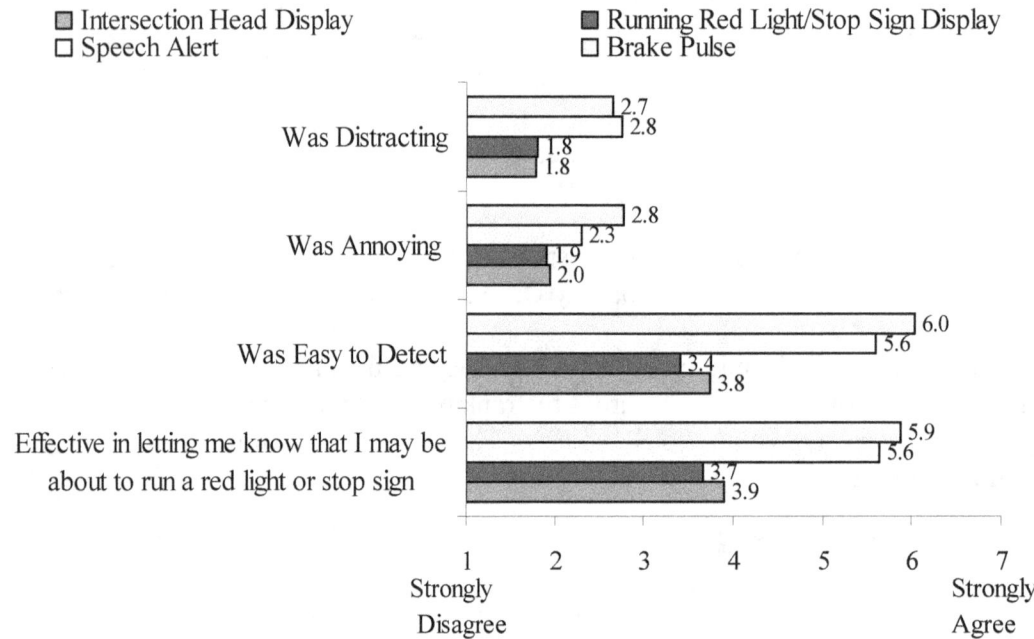

Figure 9. Mean Post-Drive Questionnaire Responses for the Usability, Distraction, and Annoyance by Alert Modality for Subjects Who Had Received at Least One CICAS-V Alert

Overall, the subjects who had received at least one alert agreed that the speech alert and the brake pulse were effective in getting their attention as shown in Figure 10. They were neutral or tended to slight disagree that the visual display was startling or effective in getting their attention. However, they agreed that the brake pulse tended to be startling and was as effective in getting their attention quickly as the speech alert.

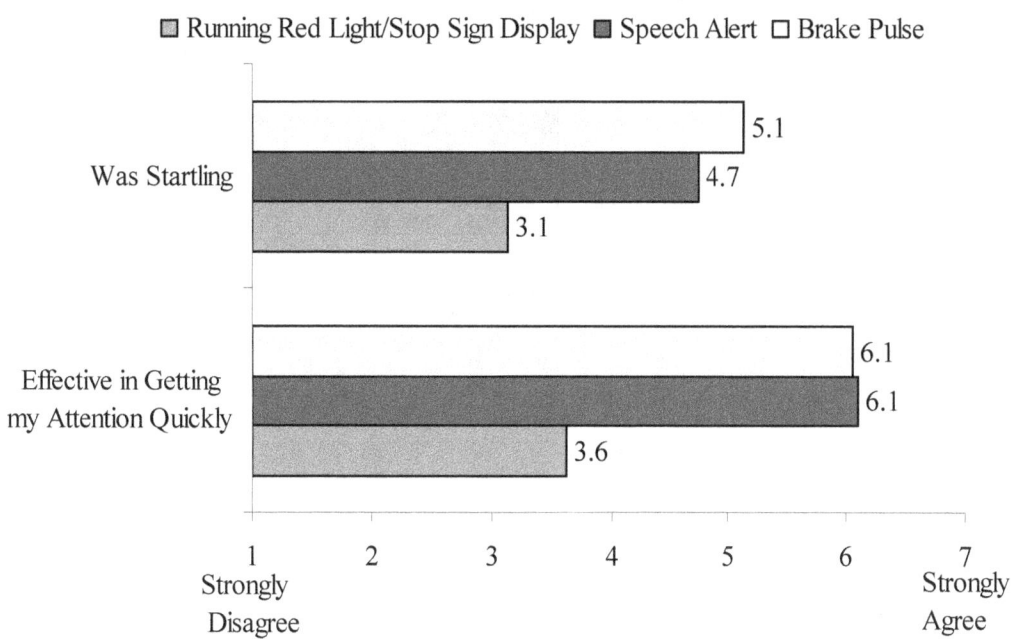

Figure 10. Mean Post-Drive Questionnaire Responses on Startling Quality and Effectiveness of CICAS-V Alerts for Subjects Who Had Received at Least One CICAS-V Alert

Figure 11 compares subjects' opinions about the usability of the brake pulse for subjects who received an alert during the pseudo-naturalistic study (21 out of the 36) with those who received an alert on the Smart Road (16 out of the 36). Both groups of subjects strongly agreed that the brake pulse was easy to detect and was effective in getting their attention quickly. Subjects who received an alert on the Smart Road strongly disagreed that the brake pulse was distracting or annoying. Those subjects in the pseudo-naturalistic study slightly disagree that the brake pulse was annoying or distracting. Subjects who received an alert during the pseudo-naturalistic driving study strongly agree that the brake pulse was startling. In contrast, subjects on the Smart Road had a neutral to very slightly agree opinion as to whether the brake pulse alert was startling. Note that most of the alerts observed during the pseudo-naturalistic study were "Running Stop Sign" alerts while all the alerts on the Smart Roads were "Running Red Light" alerts. Due to the difference in the type of alerts experienced on the Smart Road compared to the public road experience of the subjects in the pseudo-naturalistic study, it is not meaningful to compare their responses.

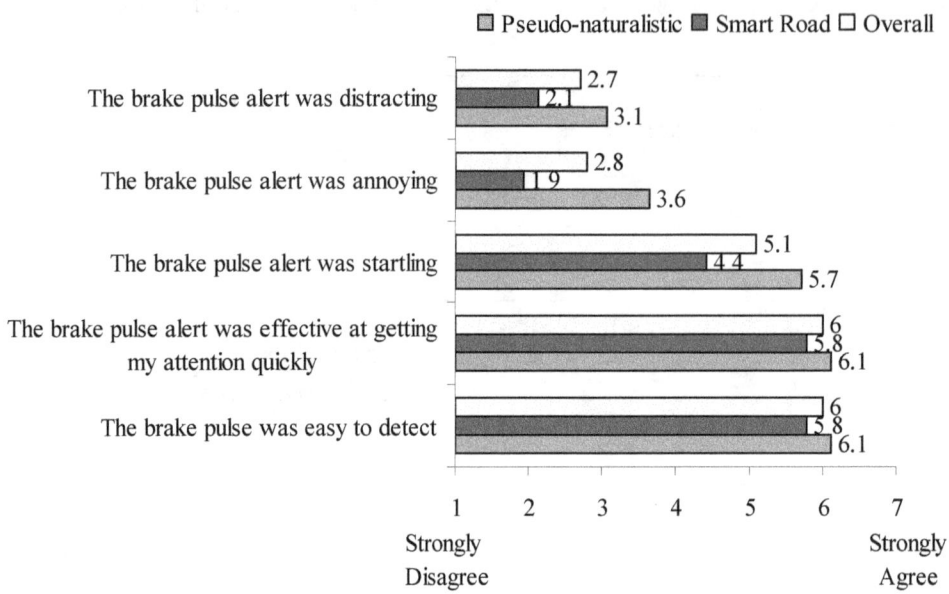

Figure 11. Responses to Questionnaire Items about Brake Pulse Usability for Subjects Who Had Received at Least One CICAS-V Alert

4.2.3. Evaluation of Subjects' Perception of Alert Timing

Subjects who received an alert (36) were asked for their opinion about the timing of the "Running Red Light" and the "Running Stop Sign" alerts. The only subjects who were asked about the "Running Stop Sign" alert timing were in the pseudo-naturalistic group. Figure 12 shows the mean scores for subject who had experienced an alert in the Smart Road and/or during the pseudo-naturalistic study. Overall, subjects rated the timing as about right and maybe a little too late. Results were similar for the "Running Red Light" and the "Running Stop Sign" alerts.

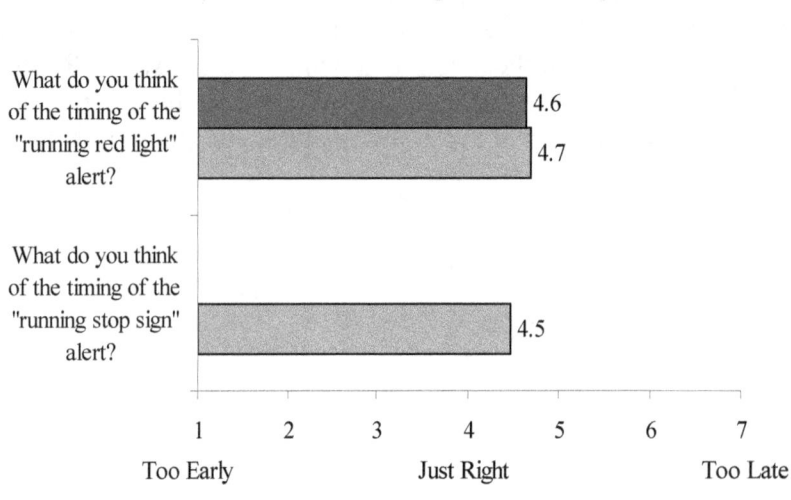

Figure 12. Responses to Questionnaire Items about Alert Timing for Pseudo-Naturalistic and Smart Road Subjects

Figure 13 shows subjects' responses to the questionnaire item asking whether they looked for following traffic when they received an alert. Subjects who received a "Running Stop Sign" alert were neutral in their response about whether they braked without checking for traffic behind them. Those who received a "Running Red Light" alert were more likely to look for traffic behind them as they applied the brakes although they remain in the neutral region.

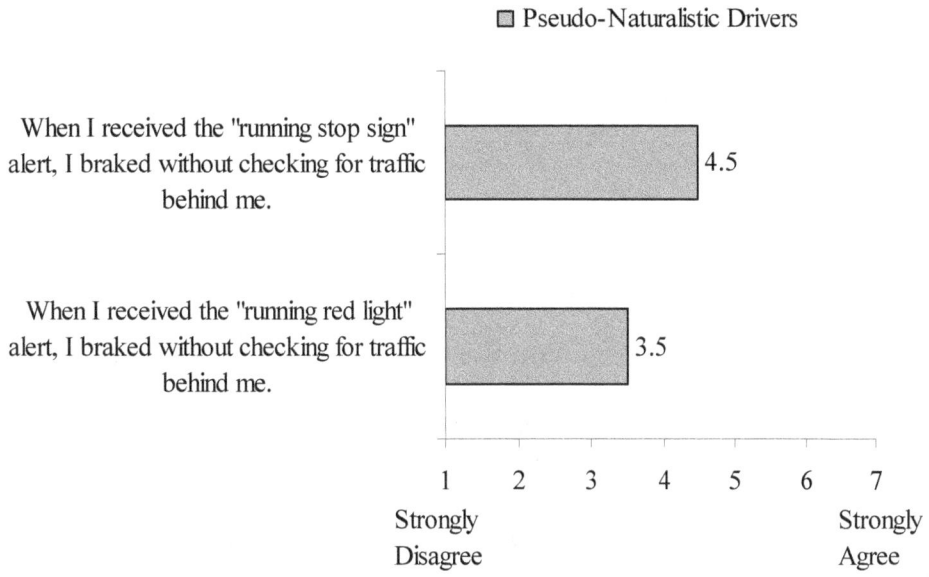

Figure 13. Responses to Questionnaire Items about Braking without Checking for Traffic for Pseudo-Naturalistic Subjects

4.2.4. Self-Reported Alert Frequency Including Nuisance Warnings

Subjects who received an alert during the pseudo-naturalistic study (21 drivers) were asked their opinions about their experience with "Running Stop Sign" alerts and "Running Red Light" alerts. Figure 14 shows the proportion of all subjects that could not identify the source of the alert, did not receive an alert when felt one was needed, received an alert they felt was not necessary, and received one they felt was appropriate. Overall, thirty-three percent of the subjects reported that they could not identify the source the alert received. Twenty-eight percent of the subjects indicated they did not receive an alert when they felt one was needed. Slightly more than four-fifths reported that they received an alert they felt was not necessary. Sixty-two percent of the subjects reported that they received a "Running Stop Sign" alert that was not necessary. The Stop Control Algorithm available to the subject may have influenced this score. As shown in Figure 14, most of the subjects reporting nuisance alerts experienced Stop Control Algorithm 1. The proportion of subjects reporting nuisance alerts looks disproportionately higher for "Running Stop Sign" alerts. However, it is important to note that there were many more subjects who received a "Running Stop Sign" alert than those receiving "Running Red Light" alerts.

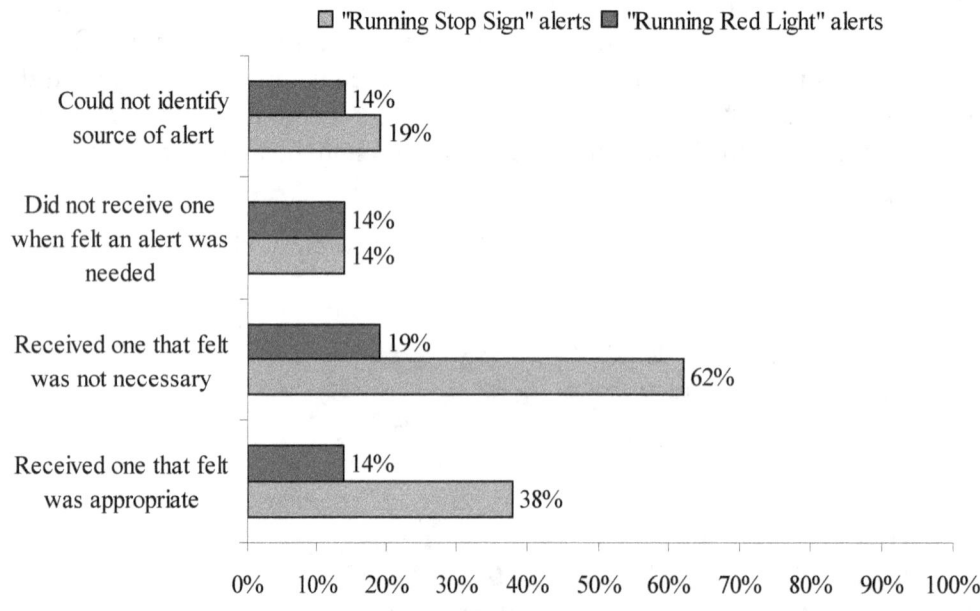

Figure 14. Responses to Questionnaire Items about Their Experience with Alerts for Pseudo-Naturalistic Subjects

Figure 15 shows the proportion of subjects who experienced at least one alert (any alert) in the pseudo-naturalistic study and could not identify the source of the alert, did not receive an alert when felt one was needed, received an alert they felt was not necessary, and received one they felt was appropriate by the stop–controlled algorithm they were exposed to. The sample size is small (14 subjects with Algorithm 1 and 7 subjects with Algorithm 2); however, the trends show that the introduction of Algorithm 2 reduced the number of subjects reporting nuisance alerts. Similarly, subjects exposed to Algorithm 2 were somewhat more likely to report that they received an alert they felt was appropriate when compared to subjects exposed to Algorithm 1.

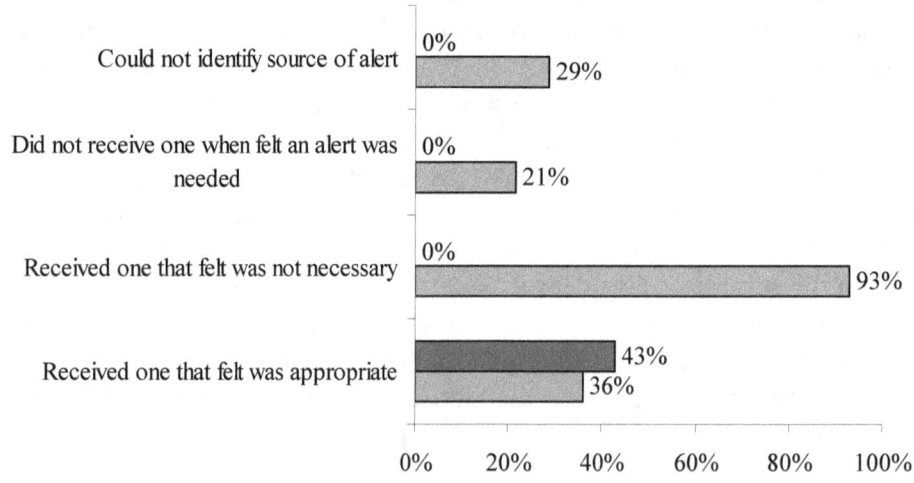

Figure 15. Responses to Questionnaire Items about Their Experience with "Running Stop Sign" Alerts for Pseudo-Naturalistic Subjects that Received at Least One Alert (any alert)

4.3. Endorsement of CICAS-V

All subjects were asked if they were interested in purchasing the CICAS-V in a new car and also to estimate how much additional money they were willing to pay to acquire this feature. Figure 16 shows that subjects were neutral to slightly negative as to whether they were willing to purchase the CICAS-V on a new vehicle. However, the data show that, on average, subjects who received at least one violation warning were neutral to slightly positive as to whether they were interested in purchasing this system. Subjects who did not receive any violation alert said they were not interested in purchasing the CICAS-V.

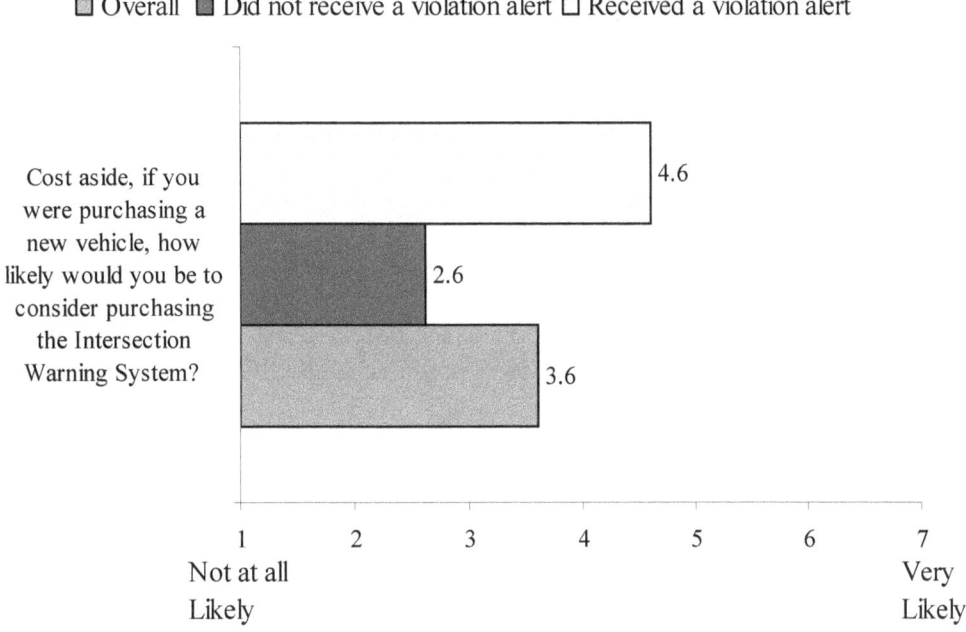

Figure 16. Post-Drive Questionnaire Responses to Inquiry about Willingness to Purchase the System by Subject Experience with CICAS-V Alerts

Subjects were also asked to estimate how much money they would pay to purchase the CICAS-V on a new vehicle. Subjects who had experienced an alert were more likely to say they would pay more to acquire the CICAS-V than those who had not received an alert (Figure 17). Overall, subjects estimated that they would pay $481 on average to acquire the CICAS-V. Subjects who had not received at least one warning said they would pay $300 on average to acquire the CICAS-V compared to $645 for subjects who had experienced an alert.[3]

---

[3] Three subjects provided outlier estimates of $10K, $20K and $35K and these were removed from the analysis.

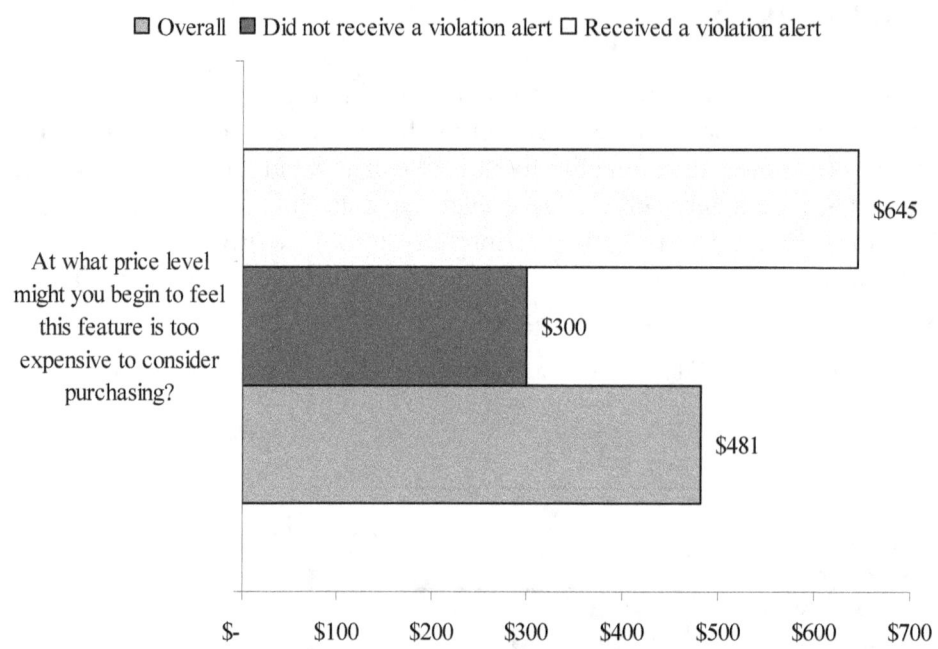

Figure 17. Post-Drive Questionnaire Responses to Inquiry as to Amount of Money Willing to Pay to Purchase by Subject Experience with CICAS-V Alerts

5. SUMMARY

The administration of the CICAS-V pilot test (i.e., subjects driving a prescribed two-hour route pilot test and a subset selected for the test track experience) as well as the expected low frequency of intersection violations limited this assessment of the acceptance of the CICAS-V system.

Overall, subjects were neutral to slightly satisfied with the CICAS-V. A similar trend was observed for their assessment of whether the system will increase their driving safety. The blue "intersection ahead" display was the only component of the CICAS-V that all the pilot test subjects experienced. Overall, the subjects had a neutral to slightly negative assessment of the "intersection ahead" display. However, when partitioned by subjects who had experienced at least one alert, the acceptance of the "intersection ahead" display increased although it still was in the neutral to slightly negative range. Subjects suggested enhancing the conspicuity of the display and using a more eye catching color or intermittent flashing to improve awareness of it.

Subjects were slightly positive overall about the safety impact of the "Running Stop Sign" and "Running Red Light" alerts. Subjects who experienced the "Running Red Light" alert when they drove on the test track were the most positive about the safety impact of the "Running Red Light" alert.

The CICAS-V implementation uses three modalities to warn subjects: visual, auditory (speech) and haptic (brake pulse). Subjects who received alerts did not rate these modalities as annoying or distracting. They rated the speech alert and brake pulse very easy to detect and effective in warning them about potential intrusion into an intersection. By contrast, subjects were neutral to slightly negative about the ease of detecting and the warning effectiveness of the "intersection ahead" display and the "running red light/stop sign" display.

Subjects were asked for their opinions of the timing of both the "Running Red Light" and "Running Stop Sign" CICAS-V alerts. They rated the timing of both of them as about right and maybe a little late.

Subjects were asked if they checked for traffic behind them when they braked after receiving the "Running Stop Sign" and "Running Red Light" alerts. They were neutral in their responses and were slightly more likely to say that they braked without checking for traffic behind them when they received the "Running Red Light" alert.

In general, between one in seven and one in five subjects said they could not identify the source of the alert and that they did not receive one when they felt it was needed. Almost two thirds of the subjects said they received a "Running Stop Sign" alert when it was not necessary which may be due to the Algorithm 1 events. Almost two in five subjects said that they received a "Running Stop Sign" alert that was appropriate.

Subjects who had received an alert slightly agreed that they would purchase the CICAS-V, cost aside. Subjects who did not receive an alert said that they definitely would not purchase the system.

Subjects' comments suggest that the CICAS-V implementation was fairly well accepted. Subjects suggested improvements to the implementation but not dismissal of the system. Among the most prominent issues they raised are the difficulty of discerning the blue color, an occasional startled reaction to the brake pulse, and a request to speed up the auditory speech message.

The data suggest that subjects need to experience the system to be able to assess it fairly. There is the possibility that the subjective definition of an intersection violation may not map exactly to the operational definition embodied in the CICAS-V. This is shown by the mismatch between the CICAS-V definition of violation versus subjective assessments. The data show that almost two thirds of the subjects reported that they thought they came close to violating a stop sign or a traffic signal but the data show that less than half or 46 percent received at least one violation alert.

6. REFERENCES

[1] Kiger, S., Neale, V., Maile, M., Kiefer, R, Ahmed-Zaid, F., Basnyake, C., Caminiti, L., Kass, S., Losh, M., Lundberg, J., Masselink, D., McGlohon, E., Mudalige, P., Pall C., Peredo, M., Stinnett, J., and VanSickle, S. *Cooperative Intersection Collision Avoidance System Limited to Stop Sign and Traffic Signal Violations (CICAS-V) Task 13 Final Report: Preparation for Field Operational Test.* Appendix J: Task 13 – Final Report: Recommended Basic FOT Design and Selection Procedures. *Cooperative Intersection Collision Avoidance System Limited to Stop Sign and Traffic Signal Violations (CICAS-V) – Phase I Final Report.* Washington, DC: Federal Highway Administration, FHWA-JPO-10-068, September 2008.

[2] Neale, V. L., Doerzaph, Z.R., Viita, D., Bowman, J., Terry, T., Bhagavathula, R., and Maile, M. *Cooperative Intersection Collision Avoidance Systems Limited to Stop Sign and Traffic Signal Violations (CICAS-V) Subtask 3.4 Interim Report: Human Factors Pilot Test of the CICAS-V.* Appendix A: Task 3 Final Report – Human Factors Development and Testing. *Cooperative Intersection Collision Avoidance System Limited to Stop Sign and Traffic Signal Violations (CICAS-V) – Phase I Final Report.* Washington, DC: Federal Highway Administration, FHWA-JPO-10-068, September 2008.

[3] Perez, M. A., Neale, V. L., Kiefer, R. J., Viita, D., Wiegand, K., and Maile, M. *Cooperative Intersection Collision Avoidance Systems Limited to Stop Sign and Traffic Signal Violations (CICAS-V) Subtask 3.3 Interim Report: Test of Alternative Driver-Vehicle Interfaces on the Smart Road.* Appendix A: Task 3 Final Report – Human Factors Development and Testing. *Cooperative Intersection Collision Avoidance System Limited to Stop Sign and Traffic Signal Violations (CICAS-V) – Phase I Final Report.* Washington, DC: Federal Highway Administration, FHWA-JPO-10-068, September 2008.

APPENDIX A. Removal and Reclassification of Some Test Data

Table 4. Data Removed from Independent Evaluation Data Set (8 Subjects)

Subject ID	Subject Experience	Volpe Comments	Action
108	SR-only	Incomplete data (as indicated by VTTI)	Responses to questionnaire items were removed from dataset
114	PN/SR	Insufficient data (completed *No Alert* questionnaire)	Responses to questionnaire items were removed from dataset
210	SR-only	Not compliant (driver looked up during surprise trial)	Responses to questionnaire items were removed from dataset
212	SR-only	Insufficient data (completed *No Alert* questionnaire)	Responses to questionnaire items were removed from dataset
304	SR-only	Incomplete data (as indicated by VTTI)	Responses to questionnaire items were removed from dataset
415	No alert	Inappropriate data (received an alert at a *non-CICAS* intersection)	Responses to questionnaire items were removed from dataset
513	SR-only	Not compliant (driver looked up during surprise trial)	Responses to questionnaire items were removed from dataset
615	SR-only	Incomplete data (as indicated by VTTI)	Responses to questionnaire items were removed from dataset

Table 5. Data Reclassified in Independent Evaluation Data Set (4 Subjects)

Subject ID	Subject Experience	Volpe Comments	Action
105	PN-only	Did not received alert; completed PN/SR questionnaire	Reclassification: "No alert". Responses to questionnaire items in sections *C, F, J, and K* were included in the data set
117	PN-only	Did not received alert; completed SR questionnaire	Reclassification: "No alert". Responses to questionnaire items in sections *B, C, G, and H* were included in the data set
309	PN-only	Did not received alert; completed PN/SR questionnaire	Reclassification: "No alert". Responses to questionnaire items in sections *C, F, J, and K* were included in the data set
516	PN-only	Did not received alert; completed SR questionnaire	Reclassification: "No alert". Responses to questionnaire items in sections *B, C, G, and H* were included in the data set

DOT HS 811 497
July 2011

U.S. Department
of Transportation
**National Highway
Traffic Safety
Administration**

www.ingramcontent.com/pod-product-compliance
Lightning Source LLC
Chambersburg PA
CBHW081806170526
45167CB00008B/3354